BIBLIOGRAPHIC RECORD TARGET

ACR0508
UL FMT B RT a BL m T/C DT 07/19/88 R/DT 07/19/88 CC STAT mm E/L 1
035/1: : |a (RLIN)MIUG86-B31333
035/2: : |a (CaOTULAS)160539667
040: : |a RPB |c RPB |d MiU
100:1 : |a Bôcher, Maxime, |d 1867-1918.
245:00: |a Regular points of linear differential equations of the second order,
|c by Maxime Bôcher.
260: : |a Cambridge [Mass.] |b Harvard University, |c 1896.
300/1: : |a 23 p. |c 22 cm.
500/1: : |a Cover-title.
650/1: 0: |a Differential equations, Linear
998: : |c RAS |s 9124

REGULAR POINTS

OF

LINEAR DIFFERENTIAL EQUATIONS

OF THE

SECOND ORDER

BY

MAXIME BÔCHER, PH.D.,
ASST. PROFESSOR OF MATHEMATICS IN HARVARD UNIVERSITY

CAMBRIDGE
Harvard University
1896

The following pages are intended quite as much for students of mathematical physics who may not be able to carry the subject farther than is here done as for those intending to make a more extended study of the modern theory of linear differential equations. Students who have mastered Byerly's *Integral Calculus* (revised edition), or its equivalent, are in a position to read what is here given. Treatises on the subject of Linear Differential Equations have recently been published by Heffter (elementary), Schlesinger and Craig. Picard's *Traité d'Analyse* and Jordan's *Course d'Analyse*, Tome III. may also be consulted with advantage.

REGULAR POINTS

OF

LINEAR DIFFERENTIAL EQUATIONS

OF THE

SECOND ORDER.

INTRODUCTION.

Any homogeneous linear differential equation of the second order may be written in the form:

$$(1) \qquad \frac{d^2y}{dx^2} + p(x)\frac{dy}{dx} + q(x)\, y = 0.$$

Two solutions of this equation are said to be linearly independent if neither is a constant multiple of the other. If y_1 and y_2 are solutions of (1) it is clear that

$$(2) \qquad y = c_1 y_1 + c_2 y_2,$$

where c_1 and c_2 are constants, will also be a solution. The converse of this is also true, namely, that if y_1 and y_2 are linearly independent *every* solution of (1) can be thrown into the form (2) which is therefore known as the general solution of (1).*

It is with equations of the form (1) that we shall have to deal in the following pages, and we will assume for the sake of simplicity that the coefficients p and q are rational functions of the independent variable x.† This variable x will be allowed to take on complex as well as real values so that we shall have to represent it geometrically by a point in the complex plane. The points, necessarily finite in number, at which at least one

* For a strict proof of this theorem cf. Heffter, pp. 45–48, 233–235.

† The reader should notice, however, that the following pages apply almost without change to the more general case in which p and q, though not rational, have no singular points except a finite number of poles in the portion of the complex plane considered.

of the functions p and q becomes infinite are called the *singular points of the differential equation*, while all other points are called non-singular or ordinary points. A justification of these terms lies in the following, —

Fundamental Theorem: If a is an ordinary point of the differential equation (1) *the general solution can be expressed in the neighborhood of this point in the form*

$$y = g_0 + g_1(x - a) + g_2(x - a)^2 + \cdot \cdot \cdot$$

where g_0 and g_1 are arbitrary constants.

We shall find it convenient to postpone the proof of this extremely important theorem until later, and to go on at once to the following,

Definition: A regular point of the equation (1) *is a point at which p does not become infinite to an order higher than the first or q to an order higher than the second.*

It will be seen that all ordinary points and some singular points are included among the regular points. The other singular points at which p becomes infinite to an order higher than the first or q to an order higher than the second or at which both of these things happen we shall naturally speak of as *irregular* points. In the following sections we shall obtain in the neighborhood of regular points solutions of (1) in the form of series and in the course of this work we shall obtain as a special case the fundamental theorem above given concerning non-singular points.

In discussing the behavior of the solutions of (1) in the neighborhood of a point a we shall find it convenient to throw the differential equation into the form:

$$(3) \quad (x - a)^2 p_0(x) \frac{d^2y}{dx^2} + (x - a)\, p_1(x) \frac{dy}{dx} + p_2(x)\, y = 0$$

where p_0, p_1, p_2 are rational functions of x none of which become infinite when $x = a$ and which do not all vanish when $x = a$. It is clear that (1) can be reduced to the form (3) in an infinite number of ways, namely by multiplying (1) through by a suitable power of $x - a$ and by any rational fraction neither the numerator nor the denominator of which contains $x - a$ as a factor. This form (3) is peculiarly convenient in distinguishing between regular and irregular points, the necessary and sufficient condition that a should be a regular point of (3) being that $p_0(a)$ shall not be zero, while the further condition that a should be a non-singular point is that p_1 should vanish at least to the first order when $x = a$ and p_2 at least to the second order.

§ 1. FORMAL TREATMENT OF REGULAR POINTS. GENERAL CASE.

Without at first making any assumption concerning the nature of the point a we will try to find a solution of (3) in the form:

$$(4) \qquad y = (x - a)^{\kappa} \sum_{\nu=0}^{\nu=\infty} g_\nu (x - a)^{\nu}.$$

We shall here assume, as we may do without loss of generality, that $g_0 \neq 0$. The above expression (4) involves an undetermined exponent κ and an infinite number of undetermined coefficients g_ν, all of which quantities must be determined by substituting (4) in the equation (3). Before performing this substitution let us note that the result of substituting $(x - a)^\rho$ in the first member of (3) is

$$(x - a)^{\rho} [\rho(\rho - 1) p_0(x) + \rho p_1(x) + p_2(x)].$$

Let the expression in square parentheses be denoted by $f(x, \rho)$. This function does not become infinite when $x = a$ and, being a rational function of x, can therefore be developed by Taylor's theorem according to powers of $x - a$:

$$(5) \qquad f(x, \rho) = \rho(\rho - 1) p_0(x) + \rho p_1(x) + p_2(x) = \sum_{\mu=0}^{\mu=\infty} f_\mu(\rho)(x - a)^{\mu}.$$

The coefficients f_μ involve ρ integrally and rationally to a degree not higher than the second.

If now we substitute (4) in the equation (3) we get:

$$\sum_{\nu=0}^{\nu=\infty} g_\nu f(x, \kappa + \nu)(x - a)^{\kappa + \nu} = 0$$

or replacing $f(x, \kappa + \nu)$ by its development (5):

$$\sum_{\nu=0}^{\nu=\infty} g_\nu \sum_{\mu=0}^{\mu=\infty} f_\mu(\kappa+\nu)(x-a)^{\kappa+\nu+\mu} = 0.$$

We must here collect the terms involving like powers of $x - a$ into a single term and equate the coefficients of the terms thus formed to zero. This gives us the following infinite set of equations:

$$(6)\quad \begin{aligned} &g_0 f_0(\kappa) = 0,\\ &g_1 f_0(\kappa+1) + g_0 f_1(\kappa) = 0,\\ &g_2 f_0(\kappa+2) + g_1 f_1(\kappa+1) + g_0 f_2(\kappa) = 0,\\ &g_3 f_0(\kappa+3) + g_2 f_1(\kappa+2) + g_1 f_2(\kappa+1) + g_0 f_3(\kappa) = 0,\\ &\cdots\cdots\cdots\cdots\cdots\cdots\cdots\cdots\\ &\cdots\cdots\cdots\cdots\cdots\cdots\cdots\cdots \end{aligned}$$

From these equations we must compute the exponent κ and the coefficients g_ν.

Since as we have seen we may assume $g_0 \neq 0$ the first equation reduces to

$$f_0(\kappa) = 0$$

an equation to determine the exponent κ and known as the *indicial equation* of the point a. Since $f_0(\kappa)$ is the constant term in the development of $f(x, \kappa)$ we have:

$$f_0(\kappa) = f(a, \kappa) = \kappa(\kappa-1)p_0(a) + \kappa p_1(a) + p_2(a).$$

The indicial equation will therefore be of the second degree unless $p_0(a) = 0$ in which case it will be of lower degree.

The necessary and sufficient condition that the indicial equation of a given point should be of the second degree is that the point should be regular.

In what follows we will assume that a is a regular point and we will denote the roots of the indicial equation, which in this case are known as the *exponents of the point* a, by κ' and κ''.

Having thus determined κ by means of the first of the equations (6) the remaining equations give us in succession the values of $g_1, g_2, g_3, \ldots$ in terms of g_0. The equations being linear the first ν of them can be solved at once by determinants and give:

(7) $g_\nu = \dfrac{(-1)^\nu h_\nu(\kappa) g_0}{f_0(\kappa+1) f_0(\kappa+2) \ldots\ldots f_0(\kappa+\nu)}$, where:

$$h_\nu(\kappa) = \begin{vmatrix} f_1(\kappa+\nu-1) & f_2(\kappa+\nu-2) & \ldots & f_{\nu-1}(\kappa+1) & f_\nu(\kappa) \\ f_0(\kappa+\nu-1) & f_1(\kappa+\nu-2) & \ldots & f_{\nu-2}(\kappa+1) & f_{\nu-1}(\kappa) \\ 0 & f_0(\kappa+\nu-2) & \ldots & f_{\nu-3}(\kappa+1) & f_{\nu-2}(\kappa) \\ 0 & 0 & \ldots & f_{\nu-4}(\kappa+1) & f_{\nu-3}(\kappa) \\ \cdot & \cdot & \cdots & \cdot & \cdot \\ \cdot & \cdot & \cdots & \cdot & \cdot \\ 0 & 0 & \ldots & f_0(\kappa+1) & f_1(\kappa) \end{vmatrix}.$$

We thus see that g_0, which occurs merely as a factor of the whole series, may be chosen at pleasure and that then $g_1, g_2, \ldots$ are completely determined the determination of course being different according as we have given to κ the value κ' or κ''. When it is necessary to indicate this last mentioned dependence of g_ν upon κ we may write $g_\nu(\kappa)$.

We have now obtained the following two series of the form (4):

$$y_1 = (x-a)^{\kappa} \sum_{\nu=0}^{\nu=\infty} g_\nu(\kappa') \, (x-a)^\nu,$$

$$y_2 = (x-a)^{\kappa''} \sum_{\nu=0}^{\nu=\infty} g_\nu(\kappa'') \, (x-a)^\nu.$$

It will be proved in § 2 that these series converge in the neighborhood of a. Assuming this it is clear that y_1 and y_2 are solutions of the differential equation (3), and, since they are in general linearly independent, that the general solution of (3) is $c_1y_1 + c_2y_2$. Two difficulties may, however, arise: 1st, the indicial equation may have equal roots $\kappa' = \kappa''$ in which case we get only one solution in the above form. 2d, some of the factors $f_0(\kappa+1)$, $f_0(\kappa+2)$, ... which occur in the denominator of $g_\nu(\kappa)$ may be zero, so that the determination of $g_\nu(\kappa)$ is impossible. This can happen only when $\kappa + n$ (n being a positive integer) is a root of the indicial equation, i. e. it can happen only when the difference of the exponents is an integer and even then only in the case of the series corresponding to the exponent with the smallest real part. We have then, except for the proof of convergency to be given later, established the following proposition:

In the neighborhood of a regular point two and only two linearly independent solutions of the differential equation can in general be found in the form (4) *corresponding to the two exponents* κ', κ'' *of the point. Exceptions occur only when* $\kappa' - \kappa''$ *is an integer in which case only one solution of the form* (4) *can in general be found.*

Owing to the fact that it is impossible to find more than two series of the form (4) which even formally (i. e. without regard to convergence) satisfy the differential equation it is obviously immaterial whether we use the above method for determining the exponents and the coefficients or any other method which will make the series satisfy the differential equation formally, the result of the determination must be the same in either case. In particular if we wish to use the above method it will make no difference how we reduce our differential equation to the standard form (3) i. e. instead of the functions p_0, p_1, p_2 we may take any other rational functions proportional to them none of which become infinite and which do not all vanish when $x = a$. We may for instance let $p_0 = 1$ as we shall find it convenient to do in § 2; or we may make p_0, p_1, p_2 all *integral* rational functions as we should usually find it convenient to do in applying the general theory to special problems since then all the coefficients $f_\mu(\rho)$ in the development of $f(x, \rho)$ after a certain point will be zero.

Finally let us glance at the case of an irregular point. Here the indicial equation is either of the zeroth or of the first degree. In the first case no solution of the form (4) exists. In the second case one and only one series of the form (4) exists which formally satisfies the differential equation, but even this series is of comparatively slight importance as, unlike the series for a regular point, it will not in general converge.*

Exercises.

1. Find the singular points of the following differential equations; distinguish between regular and irregular points and find the exponents of the regular singular points:

* For examples illustrative of the matter here touched upon see Heffter, pp. 43–44.

(a) $x^2 \frac{d^2y}{dx^2} + 2x \frac{dy}{dx} + m(m+1)y = 0.$

(b) $(1-x^2)\frac{d^2y}{dx^2} - 2x\frac{dy}{dx} + m(m+1)y = 0$

(Legendre's Equation).

(c) $(1-x^2)\frac{d^2y}{dx^2} - 2x\frac{dy}{dx} + \left[m(m+1) - \frac{n^2}{1-x^2}\right]y = 0.$

(d) $$\frac{d^2y}{dx^2} + \frac{1}{2}\left(\frac{1}{x-e_1} + \frac{1}{x-e_2} + \frac{1}{x-e_3}\right)\frac{dy}{dx}$$
$$- \frac{Ax+B}{4(x-e_1)(x-e_2)(x-e_3)}y = 0$$ (Lamé's Equation).

2. Determine the coefficients g_ν in the developments about the regular point 0 :

(a) for equation (a) Exercise 1 ;

(b) for Bessel's Equation :

$$\frac{d^2y}{dx^2} + \frac{1}{x}\frac{dy}{dx} + \left(1 - \frac{n^2}{x^2}\right)y = 0;$$

do not attempt to consider at present the cases in which the difference of the exponents is an integer.

§ 2. CONVERGENCE. GENERAL CASE.

In the preceeding section we saw that in the neighborhood of a regular point a two series of the form (4) will in general exist which formally satisfy the differential equation, while in all cases where we have a regular point one such series exists. It is the object of the present section to prove that *these series converge within a circle described about the point a as centre and passing through the nearest singular point of the differential equation.*

In the proof of this theorem we shall find it convenient to assume that $p_0 = 1$, an assumption which as we have seen is allowable.

Call the distance (real and positive) from the point a to the nearest singular point of the differential equation R. We wish to prove that the series

$$(8) \qquad \sum_{\nu=0}^{\nu=\infty} g_\nu(\kappa)\,(x-a)^\nu$$

in which the coefficients have been determined by equations (6) is convergent at any point x within a circle described about a as centre with radius R. Draw a second circle of radius K and centre at a large enough to include the point x but smaller than the first circle:

$$R > K > |\,x-a\,|\,.^*$$

We will speak of this second circle as the "circle K."

Our method of proof will be to find a set of positive real quantities $a_\nu \geqq |\,g_\nu\,|$ and to prove that the series $\Sigma\, a_\nu\,|\,x-a\,|^{\,\nu}$ is convergent. From this the absolute convergency of (8) follows. To find these quantities a_ν we proceed as follows.

From equations (6) we get:

$$g_{\nu+1} = \frac{-1}{f_0(\kappa+\nu+1)}\,[g_\nu f_1(\kappa+\nu) + g_{\nu-1} f_2(\kappa+\nu-1) + \cdot\;\cdot\;\cdot + g_0 f_{\nu+1}(\kappa)]$$

whence it follows that:

$$|\,g_{\nu+1}\,| \leqq \frac{1}{|\,f_0(\kappa+\nu+1)\,|}\,[\,|\,g_\nu\,|\cdot|\,f_1(\kappa+\nu)\,| + \cdot\;\cdot\;\cdot + |\,g_0\,|\cdot|\,f_{\nu+1}(\kappa)\,|\,].$$

Moreover we have:

$$\frac{\partial f(x,\rho)}{\partial x} = \sum_{\nu=0}^{\nu=\infty} (\nu+1) f_{\nu+1}(\rho)\,(x-a)^\nu,$$

the radius of convergence of this series being R. If then we denote the greatest value of $|\,\partial f(x,\rho)/\partial x\,|$ on the circle K by $M(\rho)$ we have by a well known theorem concerning power series:†

$$M(\rho) \geqq (\nu+1)\,|\,f_{\nu+1}(\rho)\,|\,K^\nu,$$

* By $|\,z\,|$ is meant the *absolute value* or *modulus* of the complex quantity z.

† The following theorem is refered to:

If $F(z) = c_0 + c_1 z + c_2 z^2 + \cdot\;\cdot\;\cdot$ *is a power series convergent in a circle described about the origin with radius* R *and if* $K < R$ *then the* or $$| f_{\nu+1}(\rho) | \leqq \frac{M(\rho) \cdot K^{-\nu}}{\nu+1} \leqq M(\rho) \cdot K^{-\nu},$$

therefore, —

$$|g_{\nu+1}| \leqq \frac{1}{|f_0(\kappa+\nu+1)|} [|g_\nu| \cdot M(\kappa+\nu) + |g_{\nu-1}| \cdot M(\kappa+\nu-1) \cdot K^{-1} + \cdots + |g_0| \cdot M(\kappa) \cdot K^{-\nu}].$$

The second member of this inequality we will denote by $a_{\nu+1}$ so that:

$$a_\nu = \frac{1}{|f_0(\kappa+\nu)|} [|g_{\nu-1}| \cdot M(\kappa+\nu-1) + \cdots + |g_0| \cdot M(\kappa) \cdot K^{-\nu+1}].$$

This is the quantity a_ν which we wished to define. From these expressions for $a_{\nu+1}$ and a_ν we obtain at once the equation:

$$a_{\nu+1} = \frac{a_\nu |f_0(\kappa+\nu)|}{K |f_0(\kappa+\nu+1)|} + \frac{|g_\nu| \cdot M(\kappa+\nu)}{|f_0(\kappa+\nu+1)|}.$$

In order to prove the convergency of the series $\Sigma a_\nu | x - a |^\nu$ we need merely to show that

$$\lim_{\nu=\infty} \left(\frac{a_{\nu+1}}{a_\nu} | x - a | \right) < 1.$$

Now since $| g_\nu | \leqq a_\nu$

$$\frac{a_{\nu+1}}{a_\nu} \leqq \frac{|f_0(\kappa+\nu)|}{K|f_0(\kappa+\nu+1)|} + \frac{M(\kappa+\nu)}{|f_0(\kappa+\nu+1)|}$$

and we have to find the limits of these two fractions which we will call P and Q. We evidently have:

$$f_0(\rho) = \rho^2 + A\rho + B,$$

where A and B are independent of ρ. We have then:

$$\lim_{\nu=\infty} P = \frac{1}{K} \lim_{\nu=\infty} \left| \frac{(\kappa+\nu)^2 + A(\kappa+\nu) + B}{(\kappa+\nu+1)^2 + A(\kappa+\nu+1) + B} \right| = \frac{1}{K}.$$

To compute the limit of Q let us denote by x_0 that point (or one of the points) on the circle K at which $| \partial f(x, \rho) / \partial x |$ has

greatest value which $| F(z) |$ takes on on the circle of radius K described about the origin is at least as great as $| c_\nu | K^\nu$.

For two different proofs of this theorem cf. Forsyth *Theory of Functions*, p. 33 (where however the theorem is not stated), and Harkness and Morley: *Theory of Functions*, p. 107.

the greatest value M. Then since we have assumed that $p_0 = 1$ (and it is here that this assumption is essential to our proof):

$$M(\rho) = | \rho p_1'(x_0) + p_2'(x_0) | ,$$

where the accents denote differentiation. Accordingly:

$$Q = \left| \frac{(\kappa+\nu)p_1'(x_0) + p_2'(x_0)}{(\kappa+\nu+1)^2 + A(\kappa+\nu+1) + B} \right|$$

$$\leqq \frac{|\kappa+\nu| \cdot |p_1'(x_0)| + |p_2'(x_0)|}{|(\kappa+\nu+1)^2 + A(\kappa+\nu+1) + B|} .$$

Let us denote the greatest values of $| p_1'(x) |$ and $| p_2'(x) |$ on the circle K by M_1 and M_2, then:

$$Q \leqq \frac{|\kappa+\nu| M_1 + M_2}{|(\kappa+\nu+1)^2 + A(\kappa+\nu+1) + B|} .$$

From which it follows at once that the limit as ν increases indefinitely of Q is zero. Thus we get finally:

$$\lim_{\nu=\infty} \left(\frac{a_{\nu+1}}{a_\nu} | x - a | \right) \leqq \frac{|x-a|}{K} < 1$$

and the convergency of our series is established.

We have here merely proved that the radius of convergence of the series (8) is at least as great as R, we have not proved that it is exactly equal to R and in fact in some cases it may be greater. It would not, however, be hard to show that *in general* the radius of convergence cannot exceed R.

§ 3. NON-SINGULAR AND SEMI-SINGULAR POINTS.

Among the regular points of a differential equation are included as we have seen all non-singular points. We will begin this section by attempting to apply the results so far established for regular points in general to non-singular points.

If a is a non-singular point of equation (3) the coefficients of the equation must be developable as follows, a_0 being a constant different from zero:

$$\begin{aligned} p_0(x) &= a_0 + a_1(x-a) + a_2(x-a)^2 + \cdots \\ p_1(x) &= \qquad\quad b_1(x-a) + b_2(x-a)^2 + \cdots \\ p_2(x) &= \qquad\qquad\qquad\qquad c_2(x-a)^2 + \cdots \end{aligned}$$

We have then:

$$f_0(\kappa) = \kappa(\kappa - 1)a_0$$
$$f_1(\kappa) = \kappa[(\kappa - 1)a_1 + b_1]$$
$$\cdot \quad \cdot \quad \cdot \quad \cdot \quad \cdot$$

From the first of these equations we get at once the theorem:

The exponents of every non-singular point are 0 *and* 1.

The difference of these exponents being an integer we have here one of the exceptional cases of § 1. All that we know from the preceeding sections is that corresponding to the larger of the exponents 1 a solution of the form:

$$y_1 = (x - a) \sum_{\nu=0}^{\nu=\infty} g_\nu (x - a)^\nu$$

can be found.

From the general theory given in § 1 we should expect that no series corresponding to the exponent zero would exist. If, however, in determining the coefficients g_ν of the series:

$$y = \sum_{\nu=0}^{\nu=\infty} g_\nu (x - a)^\nu$$

we use not formula (7), which as we have seen contains a zero factor in the denominator, but the equations (6) from which this formula was deduced, we see that the difficulty which usually presents itself disappears in the case we are now considering. In using equations (6) we must of course let $\kappa = 0$ so that the second of the equations (6) now becomes:

$$g_1 f_0(1) + g_0 f_1(0) = 0.$$

This equation is to be used to determine g_1, but since $f_0(1) = 0$ it appears at first sight that, since $g_0 \neq 0$, no value of g_1 will satisfy the equation. This is the difficulty above refered to which might be expected to present itself here. That it does not present itself is due to the fact that $f_1(0) = 0$ (see the formula for f_1 just given). It follows that no matter what value is given to g_1 the second of the equations (6) is satisfied, i. e. *Not only g_0 but also g_1 are entirely arbitrary in this case.*

From this point on there is no difficulty in using the equations (6) to determine in succession the coefficients $g_2, g_3, \ldots$

in terms of g_0 and g_1, and since the proof of convergence given in § 2 depends not on formula (7) but directly on equations (6) it will apply here without change.

Without actually computing g_ν in the terms of g_0 and g_1 it is clear from the form of equations (6) that we shall have: $g_\nu = g_0 g_\nu' + g_1 g_\nu''$ where g_ν' and g_ν'' are independent of the arbitrary constants g_0 and g_1. If then we assign in succession to the arbitrary constants g_0 and g_1 the values $g_0 = 1$, $g_1 = 0$ and $g_0 = 0$, $g_1 = 1$ we get the following two particular solutions:

$$y_1 = 1 + \quad * \quad + g_2'(x-a)^2 + g_3'(x-a)^3 + \cdots,$$
$$y_2 = \quad (x-a) + g_2''(x-a)^2 + g_3''(x-a)^3 + \cdots.$$

Since these two solutions are evidently linearly independent the general solution will be $y = g_0 y_1 + g_1 y_2$ which is precisely the most general series which corresponds to the exponent zero. *We have then established the fundamental theorem stated in the introduction.*

Exercises.

1. Evaluate the coefficients $g_\nu = g_0 g_\nu' + g_1 g_\nu''$ in terms of the functions $f_0, f_1, f_2, \ldots$; and show that when $g_0 = 0$ these coefficients are precisely those given by formula (7) for the series corresponding to the exponent 1.

2. Develope the general solution of Legendre's equation (§ 1, Exercise 1 (*b*)) about the non-singular point $x = 0$.

Besides the non-singular points we have just considered there are also other regular points the difference of whose exponents is an integer and yet where the difficulty which as we saw in § 1 will in this case in general present itself does not occur. Such points we will call *semi-singular points* according to the following definition:

A semi-singular point of the differential equation is a regular point the difference of whose exponents is an integer, but in the neighborhood of which, nevertheless, two linearly independent solutions of the form (4) *still exist.*

Let us find the condition that a regular point with exponents κ' and κ'', where $\kappa' - \kappa'' = n$ is a positive integer, should be a semi-singular point. It is in the determination of the coefficients of the series which corresponds to the exponent κ'' that the difficulty here presents itself, or, to be exact, in the determination of the coefficient g_n and of the subsequent coefficients. The determination of the coefficients $g_1, \ldots g_{n-1}$ in terms of g_0 presents no peculiarity. After these first coefficients have been determined the equation for determining g_n is:

$$g_n f_0(\kappa''+n) + g_{n-1} f_1(\kappa''+n-1) + \cdots + g_0 f_n(\kappa'') = 0.$$

Since $f_0(\kappa''+n) = f_0(\kappa') = 0$ the necessary and sufficient condition that there should be any value of g_n which satisfies this equation is that:

$$g_{n-1} f_1(\kappa''+n-1) + g_{n-2} f_2(\kappa''+n-2) + \cdots + g_0 f_n(\kappa'') = 0,$$

or, as is at once seen by eliminating $g_0, \ldots g_{n-1}$ between this equation and the preceding equations (6) which have been used to determine $g_1, \ldots g_{n-1}$, that: $h_n(\kappa'') = 0$ where h has the same meaning as on p. 5. If this relation is fulfilled the value of g_n is entirely arbitrary. From this point on there is no difficulty in computing the subsequent coefficients $g_{n+1}, g_{n+2}, \ldots$ in terms of g_0 and g_n.

The necessary and sufficient condition that a point, the difference of whose exponents $\kappa' - \kappa''$ is a positive integer n, should be a semi-singular point is that the determinant $h_n(\kappa'')$ should vanish.

It should be noticed that non-singular points are merely special cases of semi-singular points.

Exercises.

1. Obtain a formula similar to (7) for the coefficients $g_{n+1}, g_{n+2}, \ldots$ in the series just discussed.

2. Prove that every semi-singular point with exponents 0 and 1 is a non-singular point.

3. Can a regular point whose exponents are equal be a semi-singular point?

4. For what values of n will the point $x=0$ be a semi-singular point of Bessel's equation (§ 1, Exercise 2 (*b*))? Determine the coefficients in the development in this case.

§ 4. REGULAR POINTS. LOGARITHMIC CASE.

We come now to the case of a regular point which is not a semi-singular point, for which the difference of the exponents $\kappa' - \kappa'' = n$ is zero or a positive integer. Here one solution y_1 of the form (4) exists, namely the series corresponding to the exponent κ', but no solution of the form (4) linearly independent of y_1. In order to get a second solution we may proceed as follows. Let y_2 be any solution of (3) linearly independent of y_1 and introduce the auxiliary variable:

$$z = d\left(\frac{y_2}{y_1}\right)/dx \quad \text{or} \quad y_2 = y_1 \int z\,dx. \tag{9}$$

Substituting this last expression for y_2 in the differential equation (3) gives us when we remember that y_1 is also a solution of (3):

$$\frac{dz}{dx} + \left(2\frac{y_1'}{y_1} + \frac{p_1(x)}{p_0(x)(x-a)}\right)z = 0. \tag{10}$$

Not only does z satisfy the differential equation (10) but conversely it is easy to see that *every* solution of (10) gives when substituted in (9) a solution y_2 of (3) which except when z is identically zero is linearly independent of y_1. We can, however, solve (10) getting:

$$z = e^{-\int \left(2\frac{y_1'}{y_1} + \frac{p_1(x)}{p_0(x)(x-a)}\right)dx}.$$

Remembering now the form of y_1 we see at once that y_1'/y_1 can be expressed as a series proceeding according to ascending integral powers of $(x-a)$ and begining with the term $\kappa'(x-a)^{-1}$. Since $p_0(a) \neq 0$ the second term under the sign of integration above can also be expressed as a series proceeding according to ascending integral powers of $(x-a)$, and the first term here is $p_1(a)(x-a)^{-1}/p_0(a)$. Since κ' and κ'' are the roots of the equation

$$\kappa(\kappa-1)p_0(a)+\kappa p_1(a)+p_2(a)=0$$

we have $p_1(a)/p_0(a)=1-\kappa'-\kappa''$. Accordingly if we expand the whole expression which stands under the sign of integration above according to ascending powers of $(x-a)$ we get as the first term: $(1+\kappa'-\kappa'')(x-a)^{-1}$. Therefore

$$\int\left(2\frac{y_1'}{y_1}+\frac{p_1(x)}{p_0(x)(x-a)}\right)dx=(1+\kappa'-\kappa'')\log(x-a)+\sum_{\nu=0}^{\nu=\infty}a_\nu(x-a)^\nu.$$

Hence it follows that:

$$z=(x-a)^{\kappa''-\kappa'-1}e^{-\Sigma a_\nu(x-a)^\nu}=(x-a)^{\kappa''-\kappa'-1}\sum_{\nu=0}^{\nu=\infty}b_\nu(x-a)^\nu$$

where $b_0 \neq 0$. Since $\kappa'-\kappa''$ is zero or a positive integer there is a term in the development of z involving $(x-a)^{-1}$. We have therefore by (9):

$$y_2=y_1\left[(x-a)^{\kappa''-\kappa'}\sum_{\nu=0}^{\nu=\infty}c_\nu(x-a)^\nu+C\log(x-a)\right]$$

where $c_0 \neq 0$ except perhaps when $\kappa'=\kappa''$. Multiplying out and replacing y_1 in the first of the two terms by its development we can finally throw the above expression into the form:

$$y_2=(x-a)^{\kappa''}\sum_{\nu=0}^{\nu=\infty}g_\nu'(x-a)^\nu+Cy_1\log(x-a),$$

where $g_0' \neq 0$ except perhaps when $\kappa'=\kappa''$.

The constant C can not be zero for if it were we should have the case of a semi-singular point with which we are not at present dealing.* We may therefore simplify y_2 by dividing it through by C. Having done this we can simplify our solution still further by adding to it such a constant multiple of y_1 that the term in $(x-a)^{\kappa'}$ drops out.

* When $\kappa'=\kappa''$ the reason here given no longer holds good, but in this case C is equal to b_0 which as we saw is not zero.

In the case we are considering in this section besides the solution y_1 corresponding to the exponent κ' there is also a solution y_2 of the form:

$$(11) \qquad y_2 = \log(x-a)\cdot y_1 + (x-a)^{\kappa''} \sum_{\nu=0}^{\nu=\infty} g_\nu'(x-a)^\nu$$

where $g_n' = 0$ but (except when $n=0$) $g_0' \neq 0$.

We proceed now to the theorem:

The series in (11) *converges within a circle described about a as centre and passing through the nearest singular point of the differential equation.*

Although the truth of this theorem might be established by a careful consideration of the steps by which we deduced formula (11) it will perhaps be easier to proceed as follows: We know from the general theory of functions that the circle of convergence of any power series passes through the nearest singular point of the function developed. Suppose now that the circle within which the series of formula (11) converges (we know from the way in which the formula was deduced that it converges within *some* circle) did not reach out as far as the next singular point of the differential equation. Then it is clear that y_2 would necessarily have a singular point at some point on the circumference of the circle of convergence. This, however, would be impossible as the fundamental theorem concerning non-singular points which was stated in the Introduction shows us that y_2 being a solution of the differential equation cannot have a singular point at a non-singular point of the differential equation.

The above method although it gives us the form of y_2 does not give us any convenient means of computing the coefficients g_ν'. For this purpose we will again make use of the method of undetermined coefficients.

Let $g_{-1} = g_{-2} = \ldots = g_{-n} = 0$. We wish to satisfy equation (3) by a series of the form:

$$y_2 = \sum_{\nu=0}^{\nu=\infty} \left(g_{\nu-n}\log(x-a) + g_\nu'\right)(x-a)^{\nu+\kappa''}.$$

We saw in § 1 that the result of substituting $(x-a)^\rho$ in the first member of (3) is $(x-a)^\rho f(x,\rho)$. In the same way it is easily seen that the result of substituting $\log(x-a)\,(x-a)^\rho$ in the first member of (3) is

$$(x-a)^\rho\,[\log(x-a)\; f(x,\rho)+f'(x,\rho)]$$

where
$$f'(x,\rho)=\partial f(x,\rho)/\partial\rho.$$

If then we substitute y_2 in (3) we get:

$$\sum_{\nu=0}^{\nu=\infty}\Big[g_{\nu-n}\big(\log(x-a)f(x,\nu+\kappa'')+f'(x,\nu+\kappa'')\big)$$
$$+g_\nu' f(x,\nu+\kappa'')\Big](x-a)^{\nu+\kappa''}=0.$$

Now we have:

$$f(x,\rho)=\sum_{\mu=0}^{\mu=\infty}f_\mu(\rho)\,(x-a)^\mu \quad\text{and}\quad f'(x,\rho)=\sum_{\mu=0}^{\mu=\infty}f_\mu'(\rho)\,(x-a)^\mu$$

where the accents denote differentiation with regard to ρ.* Substitute these expansions of f and f' above and cancel out the factor $(x-a)^{\kappa''}$ and we get:

$$\log(x-a)\sum_{\nu=0}^{\nu=\infty}\sum_{\mu=0}^{\mu=\infty}g_{\nu-n}f_\mu(\nu+\kappa'')\,(x-a)^{\nu+\mu}$$
$$+\sum_{\nu=0}^{\nu=\infty}\sum_{\mu=0}^{\mu=\infty}\big(g_{\nu-n}f_\mu'(\nu+\kappa'')+g_\nu' f_\mu(\nu+\kappa'')\big)(x-a)^{\nu+\mu}=0\,.$$

It is easily seen that the series which is here multiplied into $\log(x-a)$ is identically zero; for if we collect the terms involving like powers of $(x-a)$ and remember that the g's with negative subscript are zero it is clear that the coefficients of the successive powers of $(x-a)$ are the first members of equations (6) in which κ is replaced by κ'.

We have left then only the second term of the above equation which must vanish identically. If in this series we collect the terms and equate the coefficients of the individual powers

* The development of f' need not to be obtained by differentiating the series for f term by term. We need merely to notice that:

$$f(x,\rho)=\rho(\rho-1)\,p_0(x)+\rho p_1(x)+p_2(x),$$
$$f'(x,\rho)=(2\rho-1)\,p_0(x)+p_1(x).$$

If we here expand $p_0(x)$, $p_1(x)$, $p_2(x)$ and collect terms we see at once that the coefficients in the development of f' are the derivatives of those in the development of f.

of $(x-a)$ to zero we get, remembering that not only the g's with negative subscripts but also g_n' vanish, the following infinite set of equations to determine the coefficients g_ν':

$$g_0'f_0(\kappa'')=0$$

$$g_1'f_0(\kappa''+1)+g_0'f_1(\kappa'')=0$$

(12) $$\cdot \quad \cdot \quad \cdot \quad \cdot \quad \cdot \quad \cdot$$

$$g'_{n-1}f_0(\kappa'-1)+g'_{n-2}f_1(\kappa'-2)+\cdots+g_0'f_{n-1}(\kappa'')=0$$

$$g_0f_0'(\kappa')+$$
$$* + g'_{n-1}f_1(\kappa'-1)+\cdots+g_1'f_{n-1}(\kappa''+1)+g_0'f_n(\kappa'')=0$$

$$g_1f_0'(\kappa'+1)+g_0f_1'(\kappa')$$
$$+g'_{n+1}f_0(\kappa'+1)+ * +\cdots+g_1'f_n(\kappa''+1)+g_0'f_{n+1}(\kappa'')=0$$

$$\cdot \quad \cdot \quad \cdot \quad \cdot \quad \cdot \quad \cdot \quad \cdot \quad \cdot \quad \cdot \quad \cdot \quad \cdot$$

Since $f_0(\kappa'')=0$ the first equation is fulfilled by any value of g_0' and may therefore be left out of account. By means of the 2^d, 3^d, . . . n^{th} equations we can determine $g_1', g_2' \ldots g'_{n-1}$ in terms of g_0'. The $(n+1)^{\text{st}}$ of the above equations introduces no new unknown quantity, but is to be regarded as an equation for determining in conjunction with the preceding equations the as yet undetermined quantity g_0'. Solving these n linear equations we find the value of g_0' as the ratio of two determinants of which the denominator is the expression $h_n(\kappa'')$ introduced in § 1. This expression, however, is not zero since a is not a semi-singular point (see the theorem at the end of § 3).*

Having thus determined $g_0', g_1', \ldots g'_{n-1}$ the subsequent equations allow us to determine $g'_{n+1}, g'_{n+2}, \ldots$ in succession without difficulty.

Equations (12) *allow us in all cases to determine the coefficients* g_ν'.

Having thus obtained a series of the form (11) which formally satisfies our differential equation it might seem at first sight that we ought to go on and prove that the series we have

* It is true that the above does not apply to the case in which $\kappa'=\kappa''$, but then the first $n+1$ equations which we have been considering reduce to the single equation $g_0f_0'(\kappa'')=0$ which is identically fulfilled since $f_0(\kappa)$ has $\kappa=\kappa''$ as a double root.

found converges. An explicit proof of convergency is not, however, necessary; for the considerations of the first part of this section establish the existence of a solution of the form (11) and the considerations we have just completed show that *only one* such series can even formally satisfy the differential equation. The formal solution we have just found must then be the true solution.

Exercises.

1. Prove that if we do not assign to g_n' the value zero the equations similar to (12) which we should obtain would leave g_n' entirely undetermined; and that if we assign to it a value different from zero we thereby add to our solution y_2 a constant multiple of y_1.

2. Prove that it is only when a is a regular (not semi-singular) point the difference of whose exponents is an integer that we can have a solution of the form:

$$(x-a)^{\kappa}\left(\log(x-a)\sum_{\nu=0}^{\nu=\infty} a_\nu (x-a)^\nu + \sum_{\nu=0}^{\nu=\infty} b_\nu (x-a)^\nu\right)$$

in which all the coefficients a_ν are not zero.

3. Develope about the point $x=0$ two linearly independent solutions of Bessel's equation (§ 1, Ex. 2 (b)) when n is an integer.

4. Develope about the point $x=1$ two linearly independent solutions of Legendre's equation (§ 1, Ex. 1 (b)).

§ 5. THE POINT AT INFINITY.

In the preceeding sections we have obtained solutions of our differential equation in the form of series which converge in the neighborhood of some point a of the x-plane. It is often desirable to have solutions of our equation expressed in the form of series which converge for all distant portions of the plane. Such series we shall speak of as converging in the neighborhood of the point at infinity. In order to obtain solutions in this form we will introduce the auxiliary variable

$x' = 1/x$ in the differential equation (1) which then takes on the form:

$$(13) \qquad \frac{d^2y}{dx'^2} + \frac{2x' - p(1/x')}{x'^2}\frac{dy}{dx'} + \frac{q(1/x')}{x'^4}y = 0,$$

an equation which as we see is of precisely the form (1).

Definition: The equation (1) *is said to have the point* $x = \infty$ *as a* $\left\{\begin{array}{l}\textit{non-singular}\\ \textit{semi-singular}\\ \textit{regular}\end{array}\right\}$ *point when the equation* (13) *has the point* $x' = 0$ *as such a point.*

The following theorems may be deduced at once from this definition:

The necessary and sufficient condition that $x = \infty$ *should be a regular point of* (1) *is that when* $x = \infty$ $p(x)$ *and* $q(x)$ *should vanish respectively to at least the first and second orders.*

The necessary and sufficient condition that $x = \infty$ *should be a non-singular point of* (1) *is that when* $x = \infty$ $p(x)$ *should vanish like* $2/x$ *and* $q(x)$ *should vanish at least to the fourth order.*

Finally we will give the following:

Definition: If $x = \infty$ *is a regular point of* (1) *then by the exponents of* $x = \infty$ *for the equation* (1) *are meant the exponents of* $x' = 0$ *for the equation* (13).

Suppose now that $x = \infty$ is a regular point of (1) with exponents κ' and κ''; then we can in general develope two linearly independent solutions of (13) about the point $x' = 0$ in the form:

$$(14) \qquad \begin{aligned} y_1 &= x'^{\kappa'} \sum_{\nu=0}^{\nu=\infty} g_\nu(\kappa')\, x'^{\nu}, \\ y_2 &= x'^{\kappa''} \sum_{\nu=0}^{\nu=\infty} g_\nu(\kappa'')\, x'^{\nu}. \end{aligned}$$

Now replace x' by its value $1/x$ and we get as two linearly independent solutions of (1):

$$(15) \qquad \begin{aligned} y_1 &= \left(\frac{1}{x}\right)^{\kappa'} \sum_{\nu=0}^{\nu=\infty} g_\nu(\kappa') \left(\frac{1}{x}\right)^{\nu}, \\ y_2 &= \left(\frac{1}{x}\right)^{\kappa''} \sum_{\nu=0}^{\nu=\infty} g_\nu(\kappa'') \left(\frac{1}{x}\right)^{\nu}. \end{aligned}$$

The series (14) converge when $|x'| < R$ if R is the distance from $x' = 0$ to the nearest singular point of (13). The series (15) will therefore converge when $|x| > 1/R$ i. e. the series (15) converge *outside* of a circle described about the origin with radius $1/R$. Now the form of equation (13) shows us that the singular points of (13) may be obtained from those of (1) by the transformation $x' = 1/x$. Accordingly if R is the distance from $x' = 0$ to the nearest singular point of (13) $1/R$ will be the distance from $x = 0$ to the *farthest* singular point of (1). Therefore:

The series (15) converge outside of the circle described about the origin and passing through the most distant singular point of (1).

In precisely the same way we get the following theorem:

If $x = \infty$ is a regular (not semi-singular) point the difference of whose exponents $\kappa' - \kappa''$ is a positive integer or zero two linearly independent solutions can be found in the form:

$$y_1 = \left(\frac{1}{x}\right)^{\kappa'} \sum_{\nu=0}^{\nu=\infty} g_\nu \left(\frac{1}{x}\right)^\nu,$$

$$y_2 = y_1 \log\left(\frac{1}{x}\right) + \left(\frac{1}{x}\right)^{\kappa''} \sum_{\nu=0}^{\nu=\infty} g_\nu' \left(\frac{1}{x}\right)^\nu,$$

where the series converge outside of the circle described about the origin and passing through the most distant singular point.

In any case then we see that the solutions in the neighborhood of the regular point $x = \infty$ differ in form from those in the neighborhood of a finite regular point $x = a$ only in having $(x - a)$ replaced by $1/x$. Of course the exponents and also the coefficients in the series would be obtained in any given case by actually forming equation (13) and developing its solutions about the point $x' = 0$.

Exercises.

1. For each of the differential equations in the Exercises of § 1 determine whether the point $x = \infty$ is regular or irregular, singular or non-singular; and in those cases in which it is regular determine its exponents.

2. Develope two linearly independent solutions of Legendre's equation (§ 1, Ex. 1 (*b*)) about the point $x = \infty$.

3. Prove that the most general homogeneous linear differential equation of the second order with rational coefficients whose only singular points are $e_1, e_2, \ldots e_n$ all of which are regular and have respectively the exponents κ_1', κ_1''; κ_2', κ_2''; $\ldots \kappa_n' \kappa_n''$ is

(*a*) when all the singular points are finite:

$$\frac{d^2y}{dx^2} + \left[\frac{1-\kappa_1'-\kappa_1''}{x-e_1} + \cdots + \frac{1-\kappa_n'-\kappa_n''}{x-e_n}\right]\frac{dy}{dx} + \frac{1}{f(x)}\left[\frac{\kappa_1'\kappa_1''f'(e_1)}{x-e_1} + \cdots + \frac{\kappa_n'\kappa_n''f'(e_n)}{x-e_n} + c_{n-4}x^{n-4} + c_{n-5}x^{n-5} + \cdots + c_0\right]y = 0,$$

where $f(x) = (x-e_1)(x-e_2) \ldots (x-e_n)$

and $f'(x) = df(x)/dx$.

(*b*) when $e_n = \infty$:

$$\frac{d^2y}{dx^2} + \left[\frac{1-\kappa_1'-\kappa_1''}{x-e_1} + \cdots + \frac{1-\kappa'_{n-1}-\kappa''_{n-1}}{x-e_{n-1}}\right]\frac{dy}{dx} + \frac{1}{f(x)}\left[\frac{\kappa_1'\kappa_1''f'(e_1)}{x-e_1} + \cdots + \frac{\kappa'_{n-1}\kappa''_{n-1}f'(e_{n-1})}{x-e_{n-1}} + \kappa_n'\kappa_n''x^{n-3} + c_{n-4}x^{n-4} + \cdots + c_0\right]y = 0,$$

where $f(x) = (x-e_1) \ldots (x-e_{n-1})$.

Prove that in either case we must have:

$$\kappa_1' + \kappa_1'' + \kappa_2' + \kappa_2'' + \ldots + \kappa_n' + \kappa_n'' = n-2.$$

4. The results of the last exercise apply only when $n \geqq 3$. Discuss the cases when $n < 3$.

5. Prove that if we transform the differential equation by introducing the new variable $\bar{y} = (x-a)^c y$ any finite point b will be a regular point of both differential equations or of neither and if regular will have the same exponents in both cases if $b \neq a$ while for the point $x = a$ both exponents will have been increased by c. Show also that if $x = b$ is any finite regular point whose exponents differ by an integer, b will

be a semi-singular point of both differential equations or of neither.

How will the point $x = \infty$ be affected by the transformation?

6. Prove that if we introduce the new variable $x' = \dfrac{\alpha x + \beta}{\gamma x + \delta}$ any point a of the original differential equation and the corresponding point $a' = \dfrac{\alpha a + \beta}{\gamma a + \delta}$ of the transformed equation will be either both or neither regular; and if regular will have the same exponents; and if these exponents differ by an integer will be either both or neither semi-singular.

Printed in Dunstable, United Kingdom

91228669R00020